SALVATORE MESSINA

Technical Poems

Once the threshold value is reached,
an engineer's heart beats

Translated from Italian by Luca Messina and Umberto Mattei

ISBN: 9798632621823
Publisher: Independently published

messinasalvatore@email.it

On the cover: *Racing bicycle*. National Museum of Science and Technology Leonardo da Vinci, Milan (Italy).

www.museoscienza.org

Many things happen in a dt

CONTENTS

Preface

I want to thank the reader for dedicating his precious time to a difficult reading.

A special thanks goes to those who have allowed my poems to reach you; the translation of poetry is a difficult art and, in the context of technical poems, sensitivity and technical knowledge are required. My greatest luck was to have people with these qualities by my side.
In addition to receiving the translation of this book, I received an unexpected gift, an unpublished technical poem by Luca Messina that you can find on the following page.

I hope this book can be an inspiration for the reader; should you want to share your creations, feel free to contact me.

Thanks,
Salvatore Messina

EXPONENTIAL GROWTH

Yes, it's true.
I made it in my lab.
Sabotage of an experiment
that got out of hands.
Survival of the weakest?
What a silly idea.
Monkeys will be monkeys.
Confined in their own ego,
locked down in artificial borders,
they worship their instincts
while the only true Homeland gets trashed.
They contaminate?
I sanitize.
Exponential growth,
killer inertia.
The assignment is easy, but they always fail.
Is it so hard to think forward?
Herd stupidity.
Save yourself, pray to god,
don't let the children see,
they have the right to dream?
Trust science only to save your skin,
wear a mask to cover your fear of death.
I say it 40 times:
I've had it, officially.
This sample is gone, reset and restart.
Raise the temperature of the oven,
monkeys can't swim.
But do it slowly:
so that when they see the wave coming
it will be too late.

Luca Messina

BASIC NOTIONS

TECHNOLOGY

The fingertip plastic pen

writes making noise

while a scent-free ink

takes light off a glowing sheet.

No longer by hand

but with machines

we harvest the wheat.

No longer alone

but with plows and beasts

we prepare the ground.

The fabrics,

the wheel,

the tam-tam.

STANDING STILL ON THE MÖBIUS STRIP

Feeling naked inside

when even branches dry up,

like cold metal

to caress.

Empathy with the empty world.

Eggshell.

BETWEEN BRACKETS

I am a left bracket.

We are aware pollen,

entrusted to a vector

of n random variables.

I met you

in a common realization.

You are my right bracket.

And *vice versa* by changing the sign.

OSCILLATIONS

Self-oscillating system

covered with a plastic bulrush

I swing to comfort

tired thoughts

with an easy trigger.

In my head

a stabilizing liquid

counterbalances every movement.

But what if I were out of liquid?

What if I had no head?

Metronome

I keep time for myself

knowing that the ticking will stop.

I think about it (what if I had no thoughts?).

Swing, bulrush,

swing.

A GODLY RESPONSE

You say that I refresh myself with fire

and quench my thirst with mercury,

that I am present along every dimension,

that I can divide by zero

and contain infinity.

But I have

no answer

to your question,

even though

I will have to answer.

ABSOLUTE ZERO

I won't miss the freshness,

and I won't mind the cold,

I will get chills,

I will shiver

and hardly breathe.

When the frost

lays on my skin

penetrating to the bone,

I'll watch

my body die.

Everything will become small

and absolute zero will come.

But I had hidden a +1 in the formula,

so I will stay alive somehow.

HOMO APATICUS

Running away.

What's the kick

if the door is open?

Excel.

The screen

takes my shape,

the soft keys,

an overwhelming desire

to squeeze the mouse.

Ctrl-C

Ctrl-V

The new dominant alleles

include copy-paste.

I see people

with mouths in place of ears

and I see ears

listening to themselves

while they burn in the sun

illuminating spreadsheets.

Black light

dark source

light shadows.

Convulsions,

for a peaceful acceptance

of the great passing.

Self-love

limited to intimacy only.

White fly

among the cockroaches,

tit without milk,

Sancho Panza

and Don Quixote.

Is it daytime already?

Good night.

Slavery still exists,

piracy

lynching

prostitution

Coca Cola,

give me a Coke,

religion,

give me a nun pope.

Oblivion of me

on the day of the groundhog

and of the usual gestures.

Manual for a paper-mache happiness:

this is Homo Apaticus.

CENTRIFUGAL FORCE

You grab my hands

and spin me round

lifting me

like you do with children.

And while I turn,

whirlingly,

you smile.

I'm fine.

I smile too.

You speed up and let me go.

I start to fly.

In a gentle rototranslation

I meet the ionosphere,

go past Radio Maria,

and now far from the planet

I wander.

PULSE

A fleeting glance

between two strangers,

just an instant.

Still, the fire inside:

nighttime travels

conceiving children

secret passages in a maze

new wine glasses

cheese with honey

squeezed thumbs

allergies

goodnight kisses

shoelaces ...

BETWEEN t AND t + dt

They bring me water

and throw it on the ground,

I bend over to drink

as if I were a pig.

Thank goodness no one sees me,

think of all the criticism

just because I'm thirsty.

Water becomes a net

and I find myself at sea,

the castle is now a whale

disappearing on the horizon.

I touch my forehead,

a jellyfish tells me:

"Good morning!,

if you're still thirsty

try to squeeze my body."

Jesus!

How is it possible

that only in the open sea

everybody is so helpful?

I ask about a bracelet,

I don't remember the color

just its scent inside my bed.

PROBABILITY

I bet

a hundred times

on 99 numbers out of 100.

In these hundred draws

I never won.

I don't believe in fate:

statistically unlucky.

FOR n GOING FROM 1 TO ∞

2n steps forward,

n steps backward,

a $\pi/2$ clockwise rotation

a jump on the spot

n steps to the left

n jumps forward

a $\pi/2$ clockwise rotation.

I can't stand

line dancing.

You dance.

VECTOR SUM

I create a goal.

It's new,

like a moth

I feel its power of attraction.

I decide to follow its heat.

Someone grabs my hands

softly

and puts them behind my back,

kisses my head,

handcuffs me,

tying me to their conventions.

My incredulous look

tries to drop my body

on one side,

like a wooden log

whose moment

is acted on its shorter

and less constrained side:

medicines, friends,

voices and molecules

straighten the log

gently,

leaving it vertical.

Motionless.

With the eyes open

Burned

by a light of pain.

Blind,

I'm no longer a wooden log,

I no longer have handcuffs.

I no longer feel sweetness.

Finally free

to create a new goal.

POINT

Within a point

devoid of any size

I want to live.

LAVOISIER

Affirmed

original creator,

I actually turn

existing thoughts

into new formulations.

I entangle.

I simplify.

I play.

Blamed

excellent destroyer,

I make the same thoughts

shine

in an explosion

of colours

and sighs.

Spectacular wonder.

LIMITS

One day I'll get there,

wait for me.

We won't touch

but we'll look at each other.

We'll talk.

And sniff each other.

UNIFORMLY ACCELERATED LINEAR MOTION

I don't care

about your pearl necklaces,

your expired insurance,

your house in the mountains

nor your poverty.

If we let ourselves go

we'll free fall together.

At the same speed.

BEST SELLER

He wrote more or less

a thousand books:

on principles,

on theorems

and observations.

He wrote about bridges,

fluids and mechanics,

antennas,

kinetics and diagrams.

He wrote everything,

even the system of equations

governing his life.

He never said a word.

His best seller

was an audiobook

featuring his silence.

ADVANCED NOTIONS

BALUN

Misfit,

I'm waiting for a balun

to connect me with the world

in a nuance of meanings,

in a continuum of words

towards social acceptance.

Open-minded

(as far as I guess)

I am aware that I will end up,

one way or another,

on a fifty-Ohm

commercial load.

THE THREE-STATE

Make up your mind, no time is left,

future is at the edge!

The high estate has already decided,

the signal has come

alternating with silences

even more loaded with meaning.

So much haste

in this world

of ups and downs,

disconnected

I live in the sin

of sloth: the third (e)state.

FRESNEL REFLECTIONS

He said

only

few words.

SATISFACTION MASKS

By 3 dB steps

my mood goes up and down,

looking for a level of satisfaction

suggested by a mask.

The algorithm doesn't expect

a maximum amount

of search steps.

Fallen in the loop.

Minimum tolerance.

I'm fluctuating

and I can't move on.

BETWEEN t AND t + dt #2

Thought is gone.

Like a dream

it was brilliant and true.

Thanks for interrupting me,

I was being too much of myself.

IT DEPENDS

Technical tests for freezing

electromagnetic waves

in order to observe them

with the naked eye,

caress them,

lick them,

like that pole in that cold winter.

I asked you: do you dare to do it?

You answered me

with the only possible answer:

it depends.

7Fragilecry_stal

Connected to a network,

24/7,

exposed with no active firewall,

it was easy for you thieves

to steal my identity.

"Bravo,

congratulations,

excellent work".

Reboot.

Create new profile:

Cow003254.

Enter a password.

Unsafe password.

NEAR FIELD

Way too bulky,

uncoordinated and clumsy.

And yet present,

in all modesty, indispensable.

LIFT

Not a stork,

not an eagle,

nor a reborn phoenix.

In the chicken coop,

free

inside the paddock,

the hen roams around

while pecking the ground.

On top of the henhouse

another hen

leans out

while a light breeze

goes through its comb.

It jumps off and dies.

Was it a matter of lift?

RESET

Nothing to complain about.

I eat, drink,

fulfill my needs,

the temperature is just right,

I'm not sleepy.

I work,

I share,

I'm loved

and I love.

I'm happy.

But I will never be

a fully-realized machine

until I have

the reset button.

NUCLEAR BOMB

The killer was not

the one who dropped me,

the killer was not

the one who designed me.

The killer was not

the one who gave the order,

who thought I was useful,

who carried me,

who painted me,

who gave me this shape.

I was the killer.

BETWEEN t AND t + dt #3

My luck

is that in a few minutes

I'll have forgotten everything.

And I'll be safe.

DIAGONAL

Crooked skyscrapers

and a sense of precariousness.

Far from ordinary Dolmens,

we want to create

artificial

Pisa towers.

Our flickering life is not enough,

now we want to see

the mainland hovering.

One day we'll tilt the sea too.

And we'll live diagonally,

to get there faster,

for fear of ploughing the edges.

BACKGROUND NOISE

The girl talks on the phone

with the insurance agent

who's mumbling something

in a colleague's ear.

Someone else whistles

in the car, at the traffic light.

A cat meows,

an actor clears his throat.

The train now arriving at platform 4,

the air-conditioner fan,

the crunching of the chair.

The ambulance coming to my rescue

howls in my ears.

I would like,

really,

to hear your voice.

DIESEL

Crushed

by your desires

I acknowledge

and welcome

your love.

Ready,

I let myself go.

Tired,

I give affection.

But at what price?

Toxic love.

GAUSSIANS

In a three-dimensional space

of upside-down Gaussians,

like white sheets

ready to swallow

forever,

the tightrope walker

runs,

stops,

turns around,

staggers

and stumbles.

He smiles as he falls and waits

to be welcomed by soft wadding.

TRANSFORM

The shortcut in a world

is tangled in its dual.

May Laplace one day

suggest a new way

where difficulties

can be solved

with rapid ease.

And from that day on

I will challenge to a duel

in the domain of life,

fighting

in its dual.

Once a winner

I will go back,

hoping the earned trophy

won't be my own scalp.

BETWEEN t AND t + dt #4

Like royal guards

we stand watch at the doors.

Hinge squeaking,

and few instants later

the guard disappears.

TUNNEL

Forgive me mountain

if I violate your mightiness

but your tired and fatigued body

fatigues and tires me.

And if by emptying you

I cause injuries

don't shake your mood,

don't unleash your defenses.

If one day I see the light

I will do so thanks to your wise guidance.

REYNOLDS NUMBER

Unable to stand still

I learned to let myself be dragged.

Well done myself,

like a cardboard coaster

I now travel in respectable ducts.

Faster and faster

and in even wider spaces,

I derail.

And not being able to get less dense

and not being able to flow better,

unique

I shatter in unspecified directions

melting into turbulent flow.

Farewell myself,

we were a temporary illusion.

LASER

Trapped

in a game of mirrors

I remain yet coherent

and I don't panic.

And that little bit

of me

that goes through the mirror

is lost in its usefulness.

But I,

to feel safe,

I bounce

among the thousand images

that I have of myself.

GRAETZ BRIDGE

Optimism.

Diodes smashed

with hammer blows.

The result?

Not pessimism

but the absence of output signal.

ARTIFICIAL SATELLITE

Launched into space,

latest-generation satellite,

I enjoy watching the Earth.

The part of the world I see,

illuminated for three quarters,

is silent from afar.

Shell of clouds,

pouring rain,

I can't see well.

Then I look for sun-drenched fields:

here is an expanse of wheat

brushing a small city

with its messy streets

and the scorching house roofs.

On top of a chicken coop a hen

jumps off and dies.

5 km northward a dry river

leads my gaze

to an old abandoned mill.

Consumed by love

two lovers collapse

with their sweaty fingers

still intertwined with one another.

A few steps from downtown,

a kid eats ice cream

while dangling his legs

off a green bench,

he looks up at the sky

and sticks out his tongue at me.

Launched into space,

latest-generation satellite,

I would like to pick my nose.

2^0, 2^1, 2^2, 2^3...

Waiting at 4π

its ones' complement

driven by primordial gradients.

He receives it.

Starts a dance

of protocols,

alignments,

pairings,

duplications.

Undetectable machines

building new chains.

ON and OFF messages.

Dedicated communication channels

and even more gradients.

Scrupulous instructions.

approx. $2.3E + 07$ s.

Beauty that hides away

from awareness,

you amaze me once more

revealing yourself

in the tired face of a mother

in the rough hand of a father.

BREAKWATER

Protected by a barrier

of now green breakwaters,

plods an old beach

of rough sand and seaweed.

The waves

slapping the concrete

while she waits

in slow agony

for Nature to bring her back

to the sea.

Remove the breakwaters.

Euthanasia.

CORE DRILLING

Diamond nails

swirlingly scraping

the ground.

The sound is shrill

at every stone,

at every jump.

But what little information can I get

in a few yards?

I have to dig

and hit the bottom

becoming a new research subject.

Then

another drilling,

another expedition

around the first deep hole.

Unsatisfied, I'll scrape more

widening

the deepest hole

created so far.

I will widen the pit

it will be a chasm,

and simulating a creepy sinkhole

I will create a dreadful abyss.

And then

I'll pop out the other side of the globe,

scratching the walls from the left-over rocks

until the world becomes

only a ring

for a giant finger.

All of this

in a fully peaceful state of mind

and jovial optimism.

BETWEEN t AND t + dt #5

It's raining thinking brains,

they smash on the ground

with a funny sound.

Walking in a square plaza,

I would like to be alone

me being naked

a round blunt stone.

My face slips away

and day by day

I have to ask who I am

with the help of my hands.

Shhh, the baby is asleep,

let him dream,

look at his skin so pink.

What does this silence mean?

Shhh, my body is cracking.

ALGORITHM

You could at least give me

an initial value,

I was forced

to assign you one.

If ... then

if you love me then kiss me

else ...

why did you turn your face away?

My initialization was wrong.

Trivial algorithm.

End.

ROBOT

At a distance

of 5 trillion

clock cycles,

you rest for a second.

Then you go back

to surviving.

PHOTOVOLTAIC

It's a lizard exposed to sunbeams

mild and fresh at first

then direct and blazing.

Even fish scales,

it's a photon kiss

it's electron lips.

Potential differences.

Small rivulets flow out

then streams

and tributaries

among waterfalls and parallels.

Artificial leaf

siliceous chlorophyll

Eve with no Adam

naked

gives itself to the sun.

25.4 mm

Like a screw

I try to penetrate

threaded sections

with a pitch

slightly different from mine.

Stubborn profile.

50 Hz BUZZ

I can smell the barn,

summer night,

crushed mosquitoes

on the teak furniture.

Sleeping

with a buzz

at 50 Hz.

Transformer,

are you still lulling

my relatives

in my grandparents' house?

RELATIVE

Yesterday morning

I decided to run

faster than usual:

5×10^8 m / s

I don't understand,

usually

I don't run in the morning.

EPILOGUE

ODE TO INGENUITY

Ingenuity

is a noble art,

it multiplies and divides

by equal quantities,

it decomposes and recomposes,

it exploits outstanding products

of intuitions and experiences.

Ingenuity is mutable,

in constant evolution

it lays eggs and hatches them,

it adapts and receives selection

as Nature demands

from what exists in It.

Ingenuity maximizes

efficiencies and yields,

imitates,

reuses,

confronts itself

and does not give up.

After racking his brains

the ingenious Being

suspends his work,

goes back home

and feeds on love.

Existential nourishment

building up gently,

reach the threshold value

and start making my heart beat.